Disastrous Deaths

DEATH BY FRIGHTFUL FOOD

by Mignonne Gunasekara

BEARPORT PUBLISHING

Minneapolis, Minnesota

Library of Congress Cataloging-in-Publication Data

Names: Gunasekara, Mignonne, author.
Title: Death by frightful food / by Mignonne Gunasekara.
Description: Minneapolis, MN : Bearport Publishing, [2022] | Series: Disastrous deaths | Includes bibliographical references and index.
Identifiers: LCCN 2020058669 (print) | LCCN 2020058670 (ebook) | ISBN 9781636911663 (library binding) | ISBN 9781636911717 (ebook)
Subjects: LCSH: Food–History–Miscellanea–Juvenile literature. | Food habits–Miscellanea–Juvenile literature. | Biography–Miscellanea–Juvenile literature. | Death–Miscellanea–Juvenile literature.
Classification: LCC GT2860 .G86 2021 (print) | LCC GT2860 (ebook) | DDC 394.1/2–dc23
LC record available at https://lccn.loc.gov/2020058669
LC ebook record available at https://lccn.loc.gov/2020058670

© 2022 Booklife Publishing
This edition is published by arrangement with Booklife Publishing.

North American adaptations © 2022 Bearport Publishing Company. All rights reserved. No part of this publication may be reproduced in whole or in part, stored in any retrieval system, or transmitted in any form or by any means, electronic, mechanical, photocopying, recording, or otherwise, without written permission from the publisher.

For more information, write to Bearport Publishing, 5357 Penn Avenue South, Minneapolis, MN 55419. Printed in the United States of America.

PHOTO CREDITS

All images are courtesy of Shutterstock.com, unless otherwise specified. With thanks to Getty Images, Thinkstock Photo, and iStockphoto. Background texture throughout – Abstracto. Gravestone throughout – MaryValery. Front Cover – Viktorija Reuta, Inspiring, Anatolir. 4 – pashabo, Lilkin, PixMarket. 5 – Maquiladora, Rawpixel.com, IYIKON, Krasovski Dmitri. 6 – Elegant Solution, HappyPictures, HN Works. 7 – trabantos, RRA79. 8 – Gustaf Lundberg [Public domain]*, Everett Historical. 9 – Mauritshuis [CC0], Georgios Kollidas, Kunturtle, Tayka_ya, Tomacco. 11 – Ton Reijnaerdts, NotionPic. 12 – Germany [Public domain], ottoflick, Anatolir. 13 – PHOTO FUN, COULANGES, ONYXprj. 15 – Pinkyone. 16 – Everett Historical, Djvu created by me [Public domain], ONYXprj. 17 – Everett - Art, https://commons.wikimedia.org/wiki/File Louis-Michel_van_Loo_001.jpg*. 18 – Boyko.Pictures, Orn Rin. 19 – GoodStudio, oticki. 20 – Everett Historical, Scott Latham. 21 – PFMphotostock, Tupungato, Roi and Roi, Pogorelova Olga. 22 – Lucia Fox. 23 – SofiaV, Osman Vector, Africa Studio. 24 – godrick, mejorana, Mila star. 25 – Volosina, William Hogarth [Public domain], Arisa_J. 26 – ONYXprj, MuchMania. 27 – Kevin5017. 28 – https://commons.wikimedia.org/wiki/File:Qinshihuang.jpg, Sofiaworld, MuchMania, hanukuro. 29 – Izf, Fedor Selivanov, LynxVector.
*U.S. work public domain in the U.S. for unspecified reason but presumably because it was published in the U.S. before 1924.
Additional illustrations by Jasmine Pointer.

CONTENTS

Welcome to the Disaster Zone 4
Adolf Frederick of Sweden 6
To Die For! 8
Mrs. Troffea and the
    Dancing Plague 10
Fever Food 12
Denis Diderot 14
Told You So 16
Pythagoras 18
He Must've Bean Joking 20
Henry Purcell 22
Out on a Low Note 24
Qin Shi Huangdi 26
Live Forever . . . or Don't 28

Timeline of Deaths 30
Glossary 31
Index 32
Read More 32

WELCOME TO THE DISASTER ZONE

Through the ages people have put some pretty gross and gruesome things in their mouths. So it's no surprise that some of these bad diet choices have led directly to disastrous deaths. From a harmless-looking little apricot and a bad cup of hot chocolate to a poisonous potion and an epic cream bun binge, people have long stumbled away from the dinner table directly into the grave.

Since the beginning of human history, about 107 billion people have lived on Earth. You know what that means . . . there are plenty of deaths to choose from!

In this book, we are going to look at the stories of several people who were taken out by frightful food and dreadful drinks, whether it was through overeating, **contamination**, or avoiding a field of beans.

INTO THE DISASTER ZONE WE GO . . .

Throughout history, there have been lots of sayings that mean someone has died.

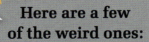
croak

Here are a few of the weird ones:

- Kicked the bucket
- Bit the dust
- Met their maker
- Six feet under
- Food for worms
- Pushing up daisies

ADOLF FREDERICK OF SWEDEN

The year was 1771, and Adolf Frederick, king of Sweden, had a bad case of the munchies.

At dinnertime, he ate lobsters, fish, caviar, and sauerkraut. He washed it all down with champagne. And that was just for his main course. Adolf had his favorite sweet treat ready for dessert—cream-filled buns. He helped himself to 14 of them! Eating all this food was not good for Adolf, and soon after this massive meal, he died of either food poisoning or **indigestion**.

Adolf Frederick was mostly a **figurehead** as king, which means he had very few powers. Sweden was **governed** by a **parliament**.

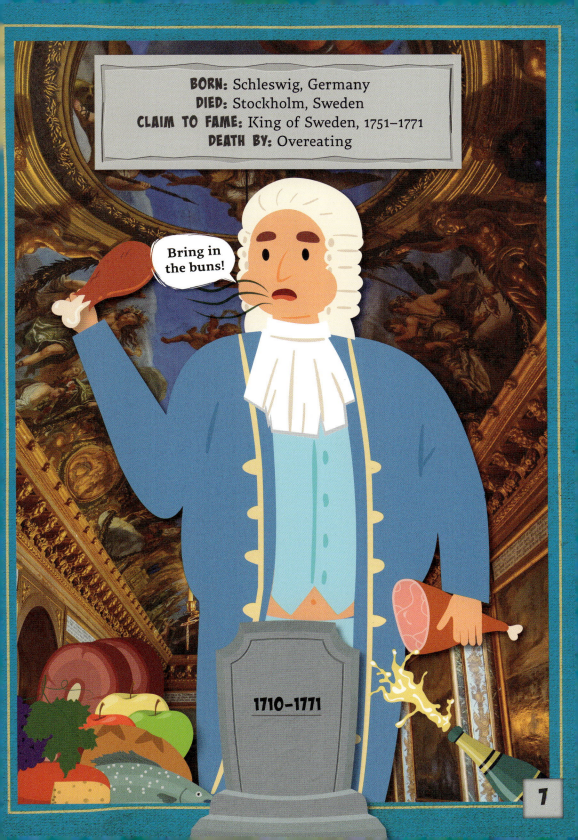

TO DIE FOR!

Like king Frederick, Queen Victoria of England also loved to eat. After her husband died, she started to eat even more to comfort herself. This made her gain a lot of weight.

Adolf Frederick of Sweden

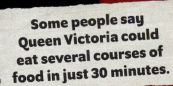

Some people say Queen Victoria could eat several courses of food in just 30 minutes.

Royals would sometimes have a whole swan or peacock served at dinner. The cooks would put the skin and feathers back on the birds after cooking them to make them look alive on the table!

These birds were **status symbols** because only the very wealthy could pay for chefs to prepare them.

Not cool.

A roast swan may have looked something like this.

Charles II was known for his love of dinner parties. Servers would offer him first choice of the food, and they would taste it to make sure it was safe.

Charles II was a king who ruled over England, Scotland, and Ireland during the seventeenth century.

Charles II liked trying new things. He was one of the first people in England to have eaten a pineapple. The first written mention of ice cream is from a menu for one of his banquets.

MRS. TROFFEA
AND THE DANCING PLAGUE

Something very strange happened in Strasbourg, France, in the summer of 1518. It began when a woman named Mrs. Troffea started to dance . . . and wouldn't stop. Within a week, she'd been joined by about 34 people.

The city's leaders decided that everyone was sick and that the only cure was to let them dance until they felt better. They cleared spaces in the city for people to dance and even hired musicians to play pipes and drums. But this only made things worse. People started to drop dead from all the dancing.

Some people say that 15 dancers collapsed every day and died from **exhaustion**, heart attacks, and strokes.

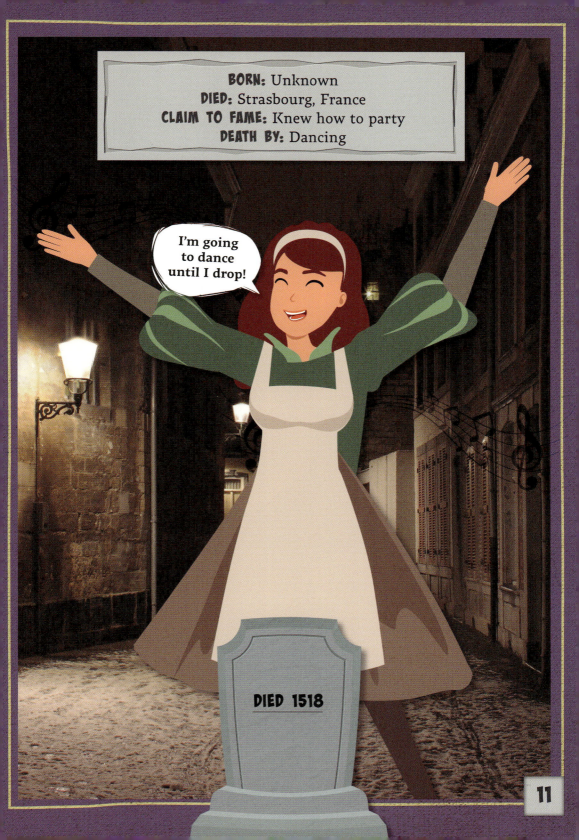

FEVER FOOD

The dancing **plague** started in mid-July and was over by September, when the city's leaders began taking the dancers to a **holy shrine,** where they seemed to be cured.

Fungus Causing the Fun

It was once thought that the Strasbourg dancers may have eaten bread or other foods that were contaminated with the fungus ergot. Eating ergot can make people see things that aren't real.

Tired dancers were held up so they could keep dancing.

Strasbourg

Strasbourg was not the only town in Europe to have suffered from a dancing plague at this time.

The ergot fungus growing on rye

Ergot poisoning would have made people sick in ways that make dancing difficult, which has some people believing this was probably not the cause of the trouble.

Ergot or Not?

The most likely explanation for the dancing plague is mass **panic**. Mass panic tends to happen in times of stress, and the people in Strasbourg were struggling with disease and **famine** at the time.

Ergot growing on wheat

13

DENIS DIDEROT

Denis Diderot was an eighteenth-century **philosopher** who wasn't afraid to question things that people thought were important or **sacred**. He also wasn't afraid to eat. One day, Denis ate a huge meal even though he was sick. As he reached for an apricot for dessert, his family warned him to stop. Denis ate the apricot anyway . . . and dropped dead soon after.

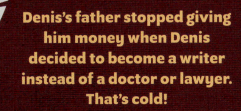

Denis's father stopped giving him money when Denis decided to become a writer instead of a doctor or lawyer. That's cold!

TOLD YOU SO

Denis Diderot was an important writer and philosopher during the Enlightenment. During this time of change people questioned the power and decisions of those in charge. They used philosophy to improve the lives of ordinary people.

Denis Diderot

Denis questioned religion in his writing. This made both the Catholic Church and the French government angry. They tried to silence Denis. When Denis kept writing anyway, they sent him to prison.

Philosophy asks questions about how people should live their lives.

16

FOR THE PEOPLE

When Denis got out of prison, he started working on an encyclopedia that he hoped would give ordinary people access to knowledge.

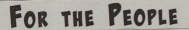

Russian Empress Catherine the Great was a big fan of Denis's work.

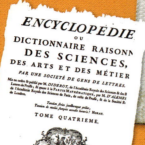

Denis wrote everything from plays to novels. When people were upset with his books, they would burn them.

"Thanks, Catherine! You really are great."

Catherine the Great helped Denis out when he needed money. She bought his library but said Denis could keep all his books until he died. She also paid him to be her librarian. Denis never had to worry about money again.

17

PYTHAGORAS

Pythagoras was a mathematician and philosopher who lived in Greece in the sixth century BCE. He had a lot of great ideas, but he and his smarty-pants students managed to anger their neighbors—a lot.

The story goes that Pythagoras was being chased by an angry mob. He was busy running for his life until he came upon a field of fava beans. Pythagoras thought fava beans were a symbol of death. He stopped running because he didn't want to go near the plants. Unfortunately, this gave the mob enough time to catch up with Pythagoras. He lost his life trying to avoid the beans!

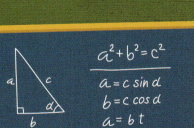

Some of the ideas Pythagoras had about life, music, and math are still important today.

BORN: Samos, Greece
DIED: Metapontum, Italy
CLAIM TO FAME: Philosopher and mathematician
DEATH BY: Avoiding beans

Oh bean, I would die for you!

AROUND 580–500 BCE

HE MUST'VE BEAN JOKING

Why did Pythagoras care that much about boring old beans? We may never know for sure because Pythagoras didn't like to write things down, and his students promised to keep what they talked about a secret.

Pythagoras and his students never ate meat and wouldn't wear animal skins or wool.

Pythagoras

A field of fava beans

Pythagoras and his followers believed that following certain strict rules was the way to live a good life. The first rule was . . . NO BEANS!

Why No Beans?

Some say Pythagoras believed that beans and humans were so similar that eating or damaging beans was like eating or hurting a human.

Fava beans are also called broad beans.

Other people say that the reason Pythagoras didn't eat beans was because they made him gassy!

The followers of Pythagoras didn't eat animals because they thought they might come back in another life as animals if they did.

A statue of Pythagoras

HENRY PURCELL

Henry Purcell is considered one of the world's greatest **composers**. He was asked to write music for the British kings and queens of the seventeenth century.

Even though he was such an important and famous man, nobody is sure how he died. One story is that he got food poisoning from some hot cocoa he drank at a chocolate house. These were cafés that served only chocolate. They were the new exciting thing in London in 1695.

Henry probably actually died from tuberculosis, an illness that affects the lungs. It had no cure during Henry's time.

OUT ON A LOW NOTE

Henry Purcell and the music he composed were very important in what is known as the Baroque period (1600–1750).

Henry Purcell

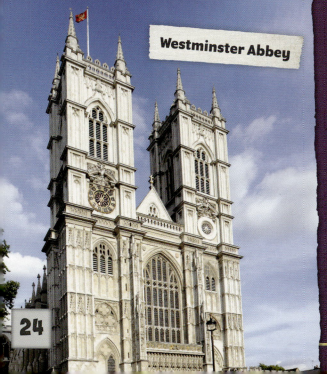

Westminster Abbey

Henry played the organ at Westminster Abbey when he was just 20 years old. He went on to compose everything from church music to music for royal birthdays. Henry is now buried in the abbey where he used to play the organ.

Henry's music is still played today.

Loco for Cocoa

Hot chocolate was quite different in the seventeenth century. It wasn't made with milk. Instead, **citrus**, vanilla, and spices, such as cinnamon and cloves, were added to it. It was an expensive drink, and only the wealthy could afford it.

Things could get pretty wild in a chocolate house.

Chocolate houses were also places to gamble, talk politics, and even plot against the king!

QIN SHI HUANGDI

In the third century BCE, Qin Shi Huangdi became the first emperor to rule over every part of China. Qin Shi Huangdi was obsessed with living forever—or at least as long as possible. He spent a lot of time and money on searching for a way to do so.

One person brought Qin Shi Huangdi a medicine that he said would make the emperor live longer if he drank it every day. The only problem was that the medicine was made with mercury, which is poisonous to humans. Qin Shi Huangdi died before he was 50 years old.

Shi Huang means first emperor.

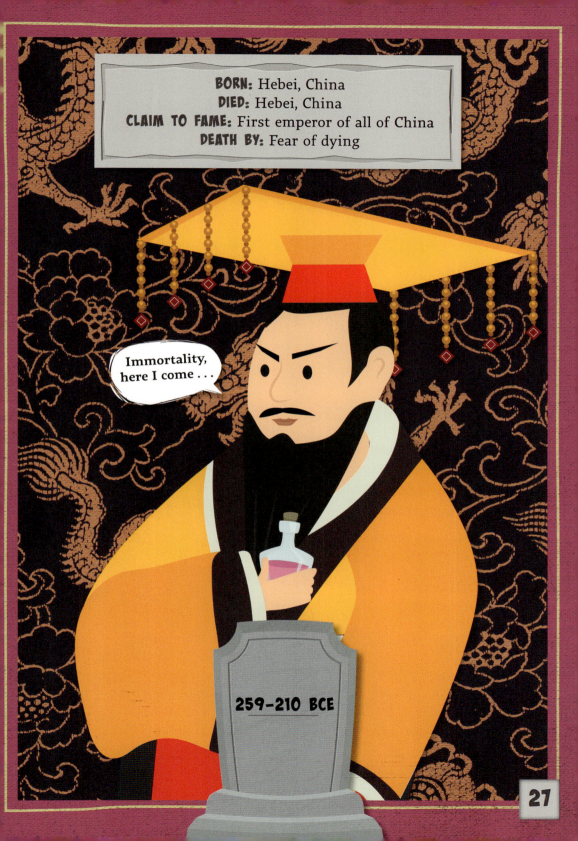

LIVE FOREVER... OR DON'T

Change It Up

When Qin Shi Huangdi became emperor, he made sure everyone followed the same laws and used the same money, measurements, and way of writing.

Qin Shi Huangdi

Qin Shi Huangdi wanted to defend China against attackers from the north, so he decided to build the Great Wall of China.

The Great Wall of China

Today, the Great Wall of China is over 13,000 miles (21,000 km) long.

Terra-Cotta Army

Qin Shi Huangdi is famous for where he was buried. He had a giant tomb built that took up 19 square miles (50 sq km)—or about half the size of Disney World! To guard this tomb, he had artists create an army of 8,000 life-sized soldiers out of terra-cotta, a type of clay.

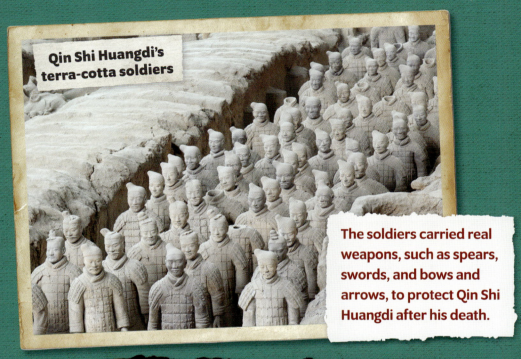

Qin Shi Huangdi's terra-cotta soldiers

The soldiers carried real weapons, such as spears, swords, and bows and arrows, to protect Qin Shi Huangdi after his death.

The terra-cotta army even had hundreds of clay horses attached to working chariots!

TIMELINE OF DEATHS

Pythagoras 500 BCE → Qin Shi Huangdi 210 BCE → Mrs. Troffea 1518

Henry Purcell 1695 → Adolf Frederick 1771 → Denis Diderot 1784

GLOSSARY

citrus the fruit of trees such as lemons, limes, oranges, and grapefruits

composers people who write music, especially as their job

contamination to make something impure by adding a poison or polluting part of it

exhaustion the state of being very tired and having little energy

famine when large numbers of people do not have enough food

figurehead a leader without real power

governed controlled or directed, especially politically

holy shrine a place or thing dedicated to a religious figure where people may pray or leave offerings

indigestion a feeling of pain or discomfort in the stomach after eating

panic extreme fear that makes it hard to act or think normally

parliament a group of people who make the laws for a country

philosopher a person who studies the nature of right and wrong, knowledge, reality, and life

plague a disease that causes death and spreads quickly to a large number of people

royals kings, queens, and members of their family

sacred highly valued or very important; very holy

status symbols things that show how rich or important someone is

INDEX

beans 5, 18–21
chocolate 4, 22–23, 25
chocolate houses 22, 25
dancing 10–13
dessert 6, 14
dinner 4, 6, 9
emperors 26–28
ideas 18
illness 10, 12–13, 22
kings 6–9, 22, 25
money 14, 17, 26, 28

music 10, 18, 22, 24
parties 9
philosophy 16
poison 4, 6, 13, 22, 26
power 6, 16
queens 8, 22
religion 16
rules 20
soldiers 29
writers 14, 16

READ MORE

Brundle, Joanna. *Famine and Drought (Transforming Earth's Geography).* New York: Kidhaven Publishing, 2018.

Croy, Anita. *Deadly Diets (The Bizarre History of Beauty).* New York: Gareth Stevens Publishing, 2019.

Goldish, Meish. *Toxic Water: Minamata, Japan (Eco-Disasters).* Minneapolis: Bearport Publishing, 2018.